A Guide to

BUSINESS CONTINUITY MANAGEMENT

Brian Doswell

Perpetuity Press Ltd
PO Box 376 Leicester LE2 1UP, UK
Telephone: +44 (0) 116 221 7778
Fax: +44 (0) 116 221 7171

214. N. Houston
Comanche Texas 76442 USA
Telephone: 915 356 7048
Fax: 915 356 3093

Email: info@perpetuitypress.co.uk
Website: http://www.perpetuitypress.com

First Published 2000

Copyright © 2000 Perpetuity Press Ltd.

All rights reserved. No part of this publication may be reprinted or reproduced or utilised in any form or by any electronic, mechanical or other means, now known or hereafter invented, including photocopying and recording, or in any information storage or retrieval system, without permission in writing from the publishers.

Warning: the doing of an unauthorised act in relation to copyright work may result in both civil claim for damages and criminal prosecution.

The views expressed in this publication are those of the author and do not necessarily reflect those of Perpetuity Press Ltd.

British Library Cataloguing in Publication Data.
A catalogue record for this book is available from the British Library.

A Guide to
Business Continuity Management
by Brian Doswell

ISBN 1 899287 57 4

ABOUT THIS GUIDE

This guide is intended to provide an introduction to business continuity management for the benefit of those who are either new to the subject or who would like to become more familiar with the scope and definitions used within the discipline.

In order to properly protect your business you must form a reasoned appreciation of the risks that might stop you trading and you must make such provision to offset those risks as you knowingly chose to afford.

This guide is intended to provide a route map for the planning process. Each of the ten steps described on the following pages is as relevant to a corner shop or a global enterprise. Step 1 requires that you address business continuity as a fully funded project rather than as an optional piece of housekeeping. Step 2 requires that you understand the risks that can affect your business and Step 3 requires that you consider the impact of those risks. Once you have understood and evaluated the risks to your business Step 4 requires that you devise and publish your corporate strategy for countering those risks should they occur. Step 5 brings in the requirements of the public emergency services as a component of your corporate strategy. Step 6 discusses the design and implementation of the business continuity plan.

Step 7 requires that the work done in forming these plans is properly communicated to those members of staff who will have a part to play in the recovery process. Step 8 seeks to ensure that the information contained in these plans is maintained in a current and useful form. Steps 9 and 10 are designed to ensure that the equally important issues of staff welfare, media relations and public authorities are not forgotten during the turmoil and disruption of the incident.

The guide includes a full glossary of terms used in the preparation of business continuity plans and ends with the internet addresses of three of the most complete sources of further information on the subject.

Contents

Section 1	Understanding Business Continuity Management	7
Section 2	Evaluation Criteria for Business Continuity Plans	13
Step 1	Project Initiation and Management	17
Step 2	Risk Evaluation and Control	19
Step 3	Business Impact Analysis	23
Step 4	Developing Business Continuity Strategies	25
Step 5	Emergency Response and Operations	27
Step 6	Developing and Implementing Business Continuity Plans	29
Step 7	Awareness and Training Programmes	37
Step 8	Maintaining and Exercising Business Continuity Plans	39
Step 9	Public Relations and Crisis Co-ordination	43
Step 10	Co-ordination with Public Authorities	47
Appendix A	General Business Continuity Terms	51
Appendix B	Additional Sources of Information	63
Acknowledgements		64

Section 1
Understanding Business Continuity Management

A BRIEF HISTORY

In truth this is not a new subject, prudent business managers have always given time and thought to what might happen if things go wrong. Most production lines have a list of 'second source suppliers' for their component parts. Most dispatch offices know where to find alternative means of delivery for their goods or services. Most administration offices have file copies of correspondence, just in case . . . . !

So what has changed to cause this aspect of general business management to rise to prominence in recent years? There are three significant answers to that question and they are:

1. The pace and volume of business in all sectors has increased many times over. Not so long ago it might have taken a day to dictate, type and send a letter. The postal service would deliver it a day or two later and the reply would take as long to return. Now that same message is dealt with in minutes via e-mail, or fax. Fewer people are able to do more work with little or no extra effort because the technology is there to help. However, when the technology breaks down the few remaining people stand little or no chance of maintaining the necessary work rate and the business is at risk. It's really a simple question of logistics.

1. UNDERSTANDING BUSINESS CONTINUITY MANAGEMENT

2. The designers of our technology acknowledge that their product is fragile in many ways and deliberately build in facilities to save and back up our valuable information files. As I write this text it is being backed up automatically by the 'system' and will be restored to me as if by magic, if or when I mess it all up. The mere fact that it is so easy to create security copies of the work done on computers, has led to a general expectation that such copies will indeed be made. Indeed a whole 'disaster recovery' industry has been built on the provision of alternative sites and alternative hardware for businesses to recover their back up copies from.

3. The third reason, and probably the most compelling, is that other people now expect it of well managed companies. Over the last five years in the UK we have seen a series of prestigious business management reports, from Cadbury, Greenbury and more recently the Turnbull Guidance on Internal Control. Each of these reports emphasise the value of information in business and urge company directors to accept a role of corporate stewardship with regard to the safe management of their shareholders' assets. Risk management is now firmly on the corporate agenda. Company directors are strongly advised to understand the risks in their business and must expect to be found guilty if they have knowledge of a problem area and have not made contingent provisions to off-set that risk. It is no longer acceptable for directors to claim that they were unaware of the risk.

THE BUSINESS CONTINUITY INSTITUTE

Accepting that business continuity planning is here to stay and that it is a part of the holistic issue of best practice for running a business, this guide sets out to define the scope of the subject using the terms published by the Business Continuity Institute (BCI). The BCI is a professional institute dedicated to providing individual practitioners with a continuous development path towards international standards of excellence in contingency planning. The body of this guide is divided into the ten subject areas that BCI identify as being significant competencies for successful practitioners. It is important to state that there is no particular significance attached to the order of the following sections, all are considered to be equally important. It is however worth noting that there is a degree of chronological sequence to the order of these ten subject areas and that many companies have used this sequence as the basis for a successful methodology for implementing their business continuity management programme.

THE DISASTER RECOVERY INSTITUTE INTERNATIONAL

For practitioners in the USA, The Disaster Recovery Institute International (DRII) offers a similar range of professional membership, education and training services. Although the BCI and the DRII are separate organisations they are both agreed on the fundamental aspects of business continuity planning and together maintain a common set of internationally accepted standards for their membership criteria. The main body of this guide is presented as a series of ten steps based on those agreed international standards. Readers will be aware from the size and presentation of this guide that there is more to say about this important aspect of corporate management. Both BCI and DRII public web sites provide a good source of additional information with detailed case studies and supporting information available to the respective institute membership.

1. Understanding Business Continuity Management

THE ROLE OF INSURANCE IN BUSINESS CONTINUITY MANAGEMENT

Corporate legislation obliges us to take conventional insurance cover for a variety of things, but we can only buy those insurance products that are available. Insurance companies will not insure everything. It is comparatively simple to insure physical corporate assets like buildings and machines where the new and/or replacement value can be precisely determined, but it is near impossible to insure logical assets such as information, market contacts and good will. Business interruption insurance is usually based on the average turnover for the previous two years, which is not always the best figure for a rapidly growing e-commerce company. In today's business environment a single PC can act as host to the entire company accounts while another similar machine may be used for little more than to produce the staff restaurant menu. In insurance terms these may be similar value assets but clearly they have differing values to the company, and that different value is not insurable.

Inevitably the most dramatic statistics come from the most dramatic incidents:

- the terrorist bomb in Bishopgate, now several years ago, is reported to have cost in the region of £800,000,000 in damage and lost business. Less than £500,000,000 was covered by insurance;
- the bomb in the World Trade Centre in New York gave rise to similar percentages of under-insurance and, perhaps typical of a litigious society, many of the insurance claims are still not settled and those companies still trading have to carry the loss.

Equally dramatic are the counter claims:

- an international clothing retailer was hit by an IRA

1. UNDERSTANDING BUSINESS CONTINUITY MANAGEMENT

bomb in the Arndale Centre in Manchester. They had contingency plans in place and were operating again within a few days of the incident. Their insurance claim was minimised by their ability to return to working operations. That company now reports that the public response to their rapid recovery resulted in an increase of total business over the accounting period.

To date the insurance industry has lacked a consistent way to evaluate the content or efficiency of contingency plans where risk means different things to different companies. This is about to change. The Loss Prevention Council (LPC), who are advisors to the Association of British Insurers (ABI), are keen to overcome this gap and are currently following the direction so well established by the quality management systems. Their approach will be to audit the process of implementing business continuity management rather than the end product, where that process includes making a formal risk assessment on which the subsequent provision of contingency measures will depend. Companies are at liberty to decide that they are not concerned about the risks or their impact and may choose to do nothing more about them, but, they must do that as a conscious decision and not by default or neglect. The LPC has chosen to describe their preferred process of contingency planning using the ten BCI subject areas as described in this guide.

INDUSTRY BEST PRACTICE

The most frequently asked question in the consulting business is 'What is everyone else doing?' True, the answer is not always quite the same thing as 'best practice' but then no one freely admits to doing anything less than their best. The Department of Trade and Industry (DTI), advisors to the private sector, and the Central Computing and Telecommunications Agency (CCTA), advisors to the public sector, both maintain Best Practice Units. Both units have publicly

1. Understanding Business Continuity Management

endorsed the approach being taken by the BCI and the LPC to evaluate business continuity management provisions.

BS7799 is currently the UK standard for information security management. As the title suggests the standard is about managing the company to provide a secure environment and business continuity management is a significant component within the scope of this standard. The text of this accreditation standard has been adopted by a growing number of countries around the world. Those countries that have not adopted BS7799 are using very similar national standards to address the overall management of information security.

Why is BS7799 so important? Because it is the only independently audited way to ascertain the management quality of a third party company when you are about to hook your network up to theirs. There are currently only a few fully qualified auditors and even fewer certified companies, but that is also changing rapidly with the impending ISO certification. Once again the auditors for BS7799 have chosen to adopt the BCI process model to evaluate the business continuity management component of the standard.

It should be noted that BS7799 is acknowledged as becoming accepted as the de facto standard for information security management throughout Europe. It is also beginning to be used as a supporting standard referenced in separate legislative areas such as the UK Data Protection Act and the European Human Rights and Freedom of Information legislation. Compliance with BS7799 is considered to be desirable as a demonstration of meeting data security requirements expressed in these information sensitive areas.

Section 2
Evaluation Criteria for Business Continuity Plans

GENERAL

In general terms a business continuity plan is intended to provide information and guidance to the corporate management team in the event of severe disruption to their normal business. The plan will not normally be invoked in the event of minor disruption where the anticipated duration of that disruption is less than that considered critical for the business affected by the disruption. It is therefore important that the plan be designed to address those factors which are, or can become, critical to the business and to ensure that the plan includes the procedures and countermeasures necessary to minimise the impact of the disruption.

The business continuity plan is clearly not meant as the blueprint for the day-to-day management of the business, but by definition, the plan must closely reflect the business which it is designed to support. The degree of support provided by the plan must be cost justified and it is customary for this justification to be a value judgement based on an estimation of the risk of disruption combined with an estimate of the impact on the business should that risk occur. Whichever way a company approaches the overall task of contingency planning this appreciation of risk and impact will influence the quality of the subsequent plan. It is also true that companies that choose to tackle the

2. Evaluation Criteria for Business Continuity Plans

task of contingency planning without completing the risk analysis first, will be forced to return to that basic question before they produce an adequate plan.

The most economical approach to the problem of business continuity planning is to follow a cyclical process comprising the following broad stages:

- business risk and impact analysis;
- corporate strategic response to the perceived risk;
- procurement and implementation of countermeasures;
- selection and training of Crisis Management Team (CMT) to operate under emergency conditions using the countermeasures provided by the plan;
- documentation of the plan to include the action plans designed to support the CMT;
- testing of the CMTs ability to operate according to plan;
- review of the test results and revision to the risk analysis and strategic plan.

The BCI would expect to see evidence of these fundamental stages in the process of constructing a comprehensive business continuity management plan.

The BCI identifies ten areas of professional competence for its members that reflect those critical steps in the planning process (See Table 1). It is inevitable that these areas of competence overlap each other in the preparation of a plan and most steps will be repeated several times during the planning exercise.

This guide explores each of the ten areas of competence to determine ways to establish the value of each area in individual plans. The leading paragraph shown in italics in each of the following sections is taken directly from the Institute's handbook.

TABLE 1
BCI – Subject Area Overview

1. **Project Initiation and Management**
   Establish the need for a Business Continuity Plan (BCP), including obtaining management support and organising and managing the project to completion within agreed upon time and budget limits.

2. **Risk Evaluation and Control**
   Determine the events and environmental surroundings that can adversely affect the organisation and its facilities with disruption as well as disaster, the damage such events can cause, and the controls needed to prevent or minimise the effects of potential loss. Provide cost-benefit analysis to justify investment in controls to mitigate risks.

3. **Business Impact Analysis**
   Identify the impacts resulting from disruptions and disaster scenarios that can affect the organisation and techniques that can be used to quantify and qualify such impacts. Establish critical functions, their recovery priorities, and inter-dependencies so that recovery time objective can be set.

4. **Developing Business Continuity Strategies**
   Determine and guide the selection of alternative business recovery operating strategies for recovery of business and information technologies within the recovery time objective, while maintaining the organisation's critical functions.

5. **Emergency Response and Operations**
   Develop and implement procedures for responding to and stabilising the situation following an incident or event, including establishing and managing an Emergency Operations Centre to be used as a command centre during the emergency.

2. Evaluation Criteria for Business Continuity Plans

6 **Developing and Implementing Business Continuity Plans**
Design, develop, and implement the Business Continuity Plan that provides recovery within the recovery time objective.

7 **Awareness and Training Programmes**
Prepare a programme to create corporate awareness and enhance the skills required to develop, implement, maintain, and execute the Business Continuity Plan.

8 **Maintaining and Exercising Business Continuity Plans**
Pre-plan and co-ordinate BCP exercises, and evaluate and document BCP exercise results. Develop processes to maintain the currency of continuity capabilities and the plan document in accordance with the organisation's strategic direction. Verify that the plan will prove effective by comparison with a suitable standard, and report results in a clear and concise manner.

9 **Public Relations and Crisis Co-ordination**
Develop, co-ordinate, evaluate, and exercise plans to handle the media during crisis situations. Develop, co-ordinate, evaluate, and exercise plans to communicate with and, as appropriate, provide trauma counselling for employees and their families, key customers, critical suppliers, owners/stockholders, and corporate management during crisis. Ensure all stakeholders are kept informed on an as-needed basis.

10 **Co-ordination with Public Authorities**
Establish applicable procedures and policies for co-ordinating response, continuity, and restoration activities with local authorities while ensuring compliance with applicable statutes or regulations.

Step 1
Project Initiation and Management

Establish the need for a Business Continuity Plan, including obtaining management support and organising and managing the project to completion within agreed upon time and budget limits.

This topic is intended to demonstrate the need for senior management commitment at the highest level to business continuity management. Business continuity management requires significant corporate commitment both in terms of time and funds if the result is to be of any value during an emergency. The management of business continuity plans should be considered as a primary budget item and managed at an appropriate level in the organisation.

Any seasoned project manager will recognise these expressions as highly desirable. If only all projects attracted the degree of top-level commitment that we would like, life would be so easy. In practice the company executives are responsible for everything that happens within their remit and, given the time, they would give top priority to everything on the list. It is important to recognise that business continuity management is naturally a second tier subject, part of the corporate housekeeping. If we make the assumption that the executives will not be against the concept of doing something to protect and enhance their company, then we can equally assume that they will favour the successful implementation of the project. Long-term project managers have often found the following tips to be very useful:

- find a project champion to keep the Board informed of your progress in the following ways:
    - make sure that the project champion knows about each successful step along the way. Concise but informal project reports on a short-term regular basis keep him /her involved;
    - encourage the project champion to report the project upwards;
    - make sure that the project champion is never surprised by bad news especially regarding costs or time-scales;
- keep a detailed project diary for reference;
- remember that it always easier to get forgiveness than it is to get permission. If in doubt, do it!

Step 2
Risk Evaluation and Control

Determine the events and environmental surroundings that can adversely affect the organisation and its facilities with disruption as well as disaster, the damage such events can cause, and the controls needed to prevent or minimise the effects of potential loss. Provide cost-benefit analysis to justify investment in controls to mitigate risks.

It is impossible to create a totally risk free environment. Every corporate organisation will be subject to its own risks, usually very closely linked to the function of the organisation and its physical location. Even the most normal of business premises can become the centre of unwelcome attention such as the McDonalds restaurant in London, recently vandalised by anti-capitalist protesters.

Fires do happen and so do floods. The UK is rightly proud of its public fire service. Each town has a 24 hour service with the whole country mapped into half-mile squares with each square plotted for risk and response to ensure that the most appropriate services are sent to a call out. Because of the quality of the fire service there are very few reported fires which burn out of control resulting in total loss. There are very few fire stations that are not called out daily to attend real fires. The London School of Media Studies was flooded from the top down when the main water tank ruptured on the roof and several thousand gallons of water found their own way down the walls, lift shafts and stairwells, causing many thousands of pounds worth of damage. Needless to say there was no flood

2. EVALUATION CRITERIA FOR BUSINESS CONTINUITY PLANS

insurance on the tenth floor of the building where they suffered the worst damage. Oddly enough, the same building suffered exactly the same problem only a few weeks later when negligent workmen allowed the new water tank to overflow while it was being installed.

There are no empirical measures for risk. There is no chart to define acceptable or unacceptable risk. The most commonly used formula is to consider threats and vulnerability together as a measure of risk as shown in Figure 1

FIGURE 1
Risk and vulnerability

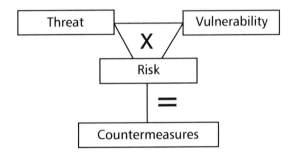

This same concept can also be shown using a scatter plot scheme based on axis representing the frequency of any event and its likely impact if and when it should occur. A typical graph is shown in Figure 2.

FIGURE 2
Scatter graph for risk assessment

Impact / Frequency of event	Very little impact	Some impact	Serious impact	Catastrophic impact
Hardly ever				
Sometimes				
Often				

Clearly both schemes rely on the user to offer an opinion. Occasionally there may be some supporting evidence such as an equipment fault log that shows how often the machine breaks down. More often the assessment will be nothing more than an impression formed by the line manager nearest to the problem. This is an area where external reference can be very useful. Third party consultants or service providers such as the fire service will usually have a broader experience of similar risks.

It is essential that all risk assessment work must have the full co-operation of the senior officers of the organisation if they are to accept the outcome of the risk assessment exercise. Ten years ago it was popular to begin presentations on this subject with the question:

'What would happen if a Boeing 747 landed on the roof of your office?'

While this question undoubtedly set the scene, it also put the whole problem beyond the realms of probability. Everyone accepts that it would theoretically be possible for

2. Evaluation Criteria for Business Continuity Plans

the 747 to crash on the roof, but for most people the likelihood of such an incident is so remote that they find it difficult to feel vulnerable to such an event. With reference to Figure 1, even a very large number multiplied by zero, equals zero.

Conversely, a smaller event such as the interruption of essential power supplies due to adjacent road works is much more believable and is likely to elicit a more positive response.

Generally most risks are of our own making in that they arise from inadequate custom and practice within the organisation.

Step 3
Business Impact Analysis

Identify the impacts resulting from disruptions and disaster scenarios that can affect the organisation and techniques that can be used to quantify and qualify such impacts. Establish critical functions, their recovery priorities, and inter-dependencies so that recovery time objective can be set.

The impact on the routine business of an organisation from the disruption arising from the occurrence of any particular risk can vary widely. It is quite possible for things to go wrong with only minimal effect on the company, or more likely that the thing that goes wrong can be fixed before the impact is felt. If a PC breaks down it is comparatively easy to get another one. If a file is lost it is comparatively easy to recover the back-up copy, assuming that there is one. If 100 PC's break down, perhaps due to a virus problem, it is still comparatively easy to repair or replace them but the time taken to achieve the repair increases.

A general rule for assessing the potential impact of any business problem is to consider how much time there is to correct the problem before it becomes critical to the adequate function of the business. If production systems have spare capacity then the loss of one line may have a minimal impact at the time. If the loss occurs at a time when the production line is full then deliveries will suffer, unless there is buffer stock in the warehouse. In other words, not all failures are fatal. In general administrative offices the pressure usually comes from the sheer volume of tasks. If the 'automated office' computer systems stop, the rate at which the backlog of work accumulates can vary with the business activity cycle.

2. Evaluation Criteria for Business Continuity Plans

The principle objective of business continuity management is to ensure that as few potential failures as possible lead to a fatal result.

It is customary for the cost justification attached to the risk control programme to depend heavily on the impact on the business of the associated risk. It is important that the impact analysis is as realistic as possible. Given that every disruption, no matter how large or small, will have some degree of negative impact on the business, it is clear that the budget to offset that impact will only be forthcoming where the senior staff of the organisation can accept the significance of the impact.

Step 4
Developing Business Continuity Strategies

Determine and guide the selection of alternative business recovery operating strategies for recovery of business and information technologies within the recovery time objective, while maintaining the organisation's critical functions.

A valid business continuity plan strategy will be one where the cost justification derived from the risk and impact analysis has been fully considered.

It is most important under the present concepts of corporate governance and good stewardship to ensure that wherever risks to the business are identified, there is an appropriate corporate response from the directors of that business stating how they intend to address the impact of the risk. There are really only three ways to deal with any given risk; remove it, accept it, or minimise the impact should it happen:

RISK REMOVAL

Risk removal is more easily said than done. The conventional approach to risk removal is to sell the risk to someone else, typically an insurance company or a third party service contractor. However, any flourishing business will experience a constant degree of change, especially in the use of information technology. Insurance will only cover some of that change as stated above. Third party suppliers may be technically and contractually responsible for the services scheduled in their contract, but compensation is not the object of the exercise.

Risk reduction through a rolling programme of review and improved resilience, is the key to minimising the residual risk. Having made an initial investment in risk management, that investment is only of value so long as the risk management programme is maintained in good order.

RISK ACCEPTANCE

The management team can always decide to accept the risk. Most companies tend to accept those risks that are either too large (such as; the proverbial aircraft landing on the roof), or too small (such as; the loss of general skilled staff where replacements can be recruited at comparatively short notice). The new interpretation of corporate stewardship says that directors should understand that these risks exist and that they should publish a risk management strategy positively stating their intention to accept the consequences. Failure to do so can be taken as ignorance of the risk or worse still neglect of responsibility.

CONTINGENCY PLANNING

Corporate stewardship decrees that if you do not accept the risk as above then you must have a contingency plan that describes how to minimise the impact on the business should the risk occur.

A comprehensive corporate risk management and business continuity strategy should address each of these three categories in sufficient detail to ensure that all affected departments understand the role that they will be expected to play should the risk occur.

Step 5
Emergency Response and Operations

Develop and implement procedures for responding to and stabilising the situation following an incident or event, including establishing and managing an Emergency Operations Centre to be used as a command centre during the emergency.

All organisations have a primary duty to their staff in the provision of emergency response plans, such as; emergency evacuation in the event of fire. These are usually tried and tested procedures that are designed to ensure that staff are cleared from the building to a nominated evacuation point. It is customary for these emergency procedures to stop at that point with no further direction as to how the business of the organisation is supported once the staff are removed to a safe location. The business continuity plan should be prepared to co-ordinate with these emergency plans and with those emergency services that may be in attendance during the emergency.

There are several case studies, especially those concerning the more dramatic terrorist bombs in London and Manchester, which include reports of looting and additional, questionable losses in the immediate aftermath of the event. While there is some truth in these reports it is also true that in most cases the problem arose primarily because of the lack of continuity between the 'Blue Light Services' (fire, police and ambulance) completing their work and the company staff taking over responsibility for the premises. It is important to understand the roles of the emergency services and to be prepared to work with them during and immediately after the event.

Step 6
Developing and Implementing Business Continuity Plans

Design, develop, and implement the Business Continuity Plan that provides recovery within the recovery time objective.

There are numerous ways to compile, store and maintain contingency plans. None are wholly right or wrong. The only significant criterion is that the end product does the job for the company that has to use it. The format for business continuity plans should match the house style of the organisation to which they apply so that the officers of that organisation recognise the document when they are required to use it, potentially under emergency conditions.

The plan should provide information and guidance to the officers who are nominated to invoke and implement the plans bearing in mind the probable circumstances under which the document is likely to be used. It is customary for this information to be presented in the form of action plans with target completion times that accord with the corporate business recovery strategy and the third party support service contracts which may have been invoked. These action plans should be supported by selected useful information such as emergency contact numbers and invocation details.

DOCUMENT FORMATS

There are well over 100 software, packaged planning products on the UK market ranging in price from £100 to £15,000 per copy. The difference in price is largely due to the amount of functionality, with the more expensive products offering complex relational database sets to hold company information while the less expensive versions are little more than word processing templates.

A word of praise and a word of caution!

The more complex the package the better quality and greater detail of information can be incorporated into the plan documents providing that it is maintained properly. If you intend to use the large database packages, you must be prepared to invest in at least one full time member of staff to collect, collate and maintain the level of detail necessary to get value for money out of the package.

All business continuity planning software packages, even the simplest of word processor templates, have the potential to act as a corporate standard for collection and presentation of the plan detail. Much of the information that is used to run the company on a good day will still be there to manage the disaster on a bad day, providing that it has been properly saved. For example, it is usually cost effective to use the company telephone book as a component of the plan rather than to create a special, crisis management, sub-set of that information which then requires additional effort to extract and maintain.

BUSINESS CONTINUITY PLAN TEMPLATES

By definition a template is a guide to the shape and content of a plan. If that guide has been prepared to suit a general market, it will require some degree of amendment in order to suit any one company. Deciding how and where to amend the template will depend on how you want to use the plan. There will inevitably be a certain amount of iteration as the plans are customised and refined and in order to minimise the number of iterations it is wise to consider the following approach:

- examine the business continuity plan strategy closely to determine the business priority for the allocation of scarce or skilled resources during the emergency;
- determine where those skilled or scarce resources are going to come from;
- nominate the staff or contractors who will participate in the planned activity;
- amend the template to suit the nominated people who must understand the requirements that the plan places on them;
- with luck the business continuity plan document will only ever be used to support test sessions. Do not expect the users to be totally familiar with the content. Do not be afraid to use jargon, technical terms or common use expressions as appropriate to ensure that the users of the document will understand what is required of them.

2. EVALUATION CRITERIA FOR BUSINESS CONTINUITY PLANS

WHO TO INVOLVE

It is common to find that the business continuity plan has its origins in the IT department of the company as a reflection of the importance of IT services in any modern business. However, when disaster strikes, everyone will be involved, from the most senior executive to the lowest clerical assistant. The business continuity plan should not be expected to nominate detailed roles for everyone in the company but it must acknowledge those people who have special knowledge or responsibility. For example:

- the chief executive officer and the directors have the ultimate responsibility for the company. They may choose to delegate the preparation of the plan, but they will be critical to its execution on the day. The directors will be required to make the overall strategic decisions and they must be included in the executive action aspects of the plan. For example:

    - media relations;
    - staff welfare;
    - authorisation of emergency funding;
    - arbitration over the allocation of scarce resources;
    - future business strategy in the aftermath of the event;

- business unit managers are the most likely to understand exactly what has gone wrong and what the priority of the day is within their individual business areas. They must identify the necessary work round procedures appropriate to recover from the actual damage of the event. Each business unit manager should have a customised section of the plan to support that particular business function with a positive indication of the target times for recovery of the essential support services;

- service unit managers are those who provide a service such as IT, facilities, personnel or training, to the business units. These are usually the owners/managers of third party support contracts, typically the IT department may have an off site recovery centre to back-up their computer systems. The service unit business continuity plan must contain the appropriate invocation instructions for the third party contracts and the target times for the re-construction of essential services to the business units;
- service units are also business units albeit their customers are internal to the company. The first task of the service unit at this time will be to support the business but service unit managers should remember that they will also have corporate business responsibilities to recover in parallel with their support function.

TYPICAL PLAN CONTENTS

A typical business continuity plan action plan will identify a phased response to crisis management. Bearing in mind that the plan is primarily designed to support managers during the first few hours and days following a major incident it is usual to find plans divided into three major phases:

- initial response;
- controlling the disruption;
- recovering from the disruption.

A summary of the actions to be performed within each phase and a business continuity plan flowchart are shown in Table 2 and Figure 3.

2. Evaluation Criteria for Business Continuity Plans

TABLE 2
A Summary of the actions to be performed within each phase

Recovery phase	Action	Reference
Initial response	Emergency invocation	Section 1
	Telephone tree	Section 2
	What to do	Section 3
	Where to go	Section 4
Controlling the disruption	Business unit meeting - problem analysis	Section 5
	Damage assessment	Section 6
	Recovery requirements definition	Section 7
	Emergency communications	Section 8
Recovering from the disruption	Continuing actions	Sections 9
	Business resumption – return to normal operation	Section 10

A typical plan will also include a number of appendices containing useful information in support of the above sections and available to all plan users as indicated in the list below:

- Appendix A: Contacts
- Appendix B: Incident and damage reports
- Appendix C: Management team control centre
- Appendix D: Team leaders meeting
- Appendix E: Recovery time estimate
- Appendix F: Information and reporting
- Appendix G: Alert and invocation
- Appendix H: Business continuity activity logs
- Appendix I: Software inventory
- Appendix J: Computer capacity requirements
- Appendix K: Backup communication network
- Appendix L: Material in fireproof safes & off-site storage
- Appendix M: Log of changes to business continuity plan

The appendices may also contain the details of third party standby contracts, especially those requiring complex instructions for the teams receiving the contracted service.

2. Evaluation Criteria for Business Continuity Plans

FIGURE 3
Business continuity plan flowchart

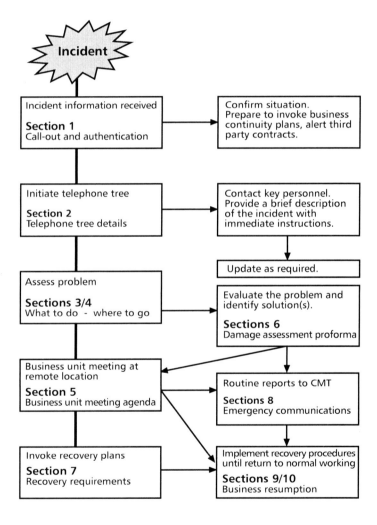

IMPLEMENTING THE PLAN

Implementation is not a single function to be achieved as a milestone on a project plan. Implementation of business continuity management is similar to the implementation of a quality management system that requires as much cultural change as it does contract placement. Planning is an essential step; back-up copies of data files are essential; testing and maintenance are essential, but unless the people involved in the day to day business accept that they must participate in the security of their own working functions, the plan will never fully deliver the benefits expected of it.

Most business continuity management programmes will have some definitive milestones, such as the agreement to stand-by contracts or the commencement of an off-site storage service. The issues of plan documents to all business units are often only the tangible aspects of the programme. Unfortunately the greater part of business continuity management is not easily visible unless a special effort is made to explain the strategy and the background work required to keep it all in place.

Step 7
Awareness and Training Programmes

Prepare a programme to create corporate awareness and enhance the skills required to develop, implement, maintain, and execute the Business Continuity Plan.

While most business managers will be competent in their primary role, it is likely that very few will have experienced the need to respond to a severe disruption to their working environment. It is important that those officers who are nominated to respond within the scope of the business continuity plans are given the opportunity to understand the rationale for their role in the emergency and to explore how best to rehearse their response to the plan.

Business continuity management should be understood and accepted as being a 'housekeeping' issue. It is an important subject but not necessarily more important than many other 'best practice' topics that underpin a successful business. Business continuity management suffers more than most from this second tier existence because, like other insurance policies, there is often very little evidence of its implementation. Most companies are able to show some evidence of their corporate insurance by exhibiting a certificate in the reception area. If business continuity management is to get similar treatment it well worth investing in a similar public flag even if it is self-generated.

'If you've got it, flaunt it!'

A more practical way to leave a trail of evidence is to make a point of marking high value assets and secure cupboards

2. Evaluation Criteria for Business Continuity Plans

to remind staff to treat them accordingly. If you decide to keep valuable contract documents in a fireproof cupboard, an elaborate identifying marker will remind staff that the procedure is there to be used and it will help the salvage team to recover your vital documents more quickly should it be necessary.

In most cases the initial approach to awareness and training is limited to those who have a need to know. The nominees who have a role to play in the planning and response activity will usually receive some degree of briefing during the planning process. However, when the incident occurs all members of staff will need to know what is expected of them after they have ... *'assembled at the muster point'*. It is strongly recommended that all members of staff are made aware of the general provisions for business continuity management through the company newsletter or notice board announcements. New joiners should receive a briefing as part of their induction process. Remember, saving the company is saving their jobs.

Step 8
Maintaining and Exercising Business Continuity Plans

Pre-plan and co-ordinate BCP exercises, and evaluate and document BCP exercise results. Develop processes to maintain the currency of continuity capabilities and the plan document in accordance with the organisation's strategic direction. Verify that the plan will prove effective by comparison with a suitable standard, and report results in a clear and concise manner.

Most plans will comprise two elements:

- the provision of various 'standby' goods and services at short notice;
- the ways in which these emergency supplies are intended to be used to effect the recovery and continuity of the business.

Business continuity plans are usually vulnerable in three ways:

- the maintenance of the scope and provision within the plan as compared with the growth and changes within the business which the plan is designed to support;
- the goods and services provided within the plan may not work as expected;

2. EVALUATION CRITERIA FOR BUSINESS CONTINUITY PLANS

- the staff nominated to participate in the recovery process may not be fully conversant with their roles.

The exercising and maintenance of these plans should show how these topics will be reviewed on a regular basis and how the shortfalls will be recorded and corrective action applied.

BUSINESS CONTINUITY PLAN TESTING

Testing is an expression that implies a pass or fail result. Testing is fine if the item under test is a technical box or procedure that may or may not work according to expectation. Testing is less acceptable when it implies a performance measure that has the potential to reflect adversely on the people concerned. If the business continuity plan strategy requires a replacement computer system to be available within a given time, it is important to prove that the measures put in place to achieve that result do work as expected. It is accepted that the people who are required to exercise that countermeasure will become part of the 'test', but the emphasis should still be on the tools being available to them rather than their ability to use them correctly.

Testing is especially important when third party contracts are an essential part of the plans and the ability to control or influence the contracted service is inherently outside the company control. If the third party service is not tested on a regular basis, the corporate business continuity management team will never fully appreciate what they have and what to do with it.

BUSINESS CONTINUITY PLAN REHEARSALS

The crisis management business continuity management teams will be drawn from the senior managers of the company who often have very busy lives. Business continuity management will not be high on their bedtime reading list. When these people are invited to participate in a business continuity management exercise most of them will not have thought about the topic for some time. The objective of the exercise must be to refresh their understanding of their role in the plans and give them an opportunity to experience some of the more unusual demands that might be made on them under emergency conditions.

In many ways a business continuity management exercise will be like an amateur dramatic society rehearsing a script with a story line that has very little to do with their normal professional lives. Every rehearsal will improve their collective potential to put on a good show. Every rehearsal will enable individuals to practice their own role and get a better understanding of the problems that their colleagues will be experiencing. Every rehearsal will be a learning experience and should lead to a wish list of things to improve the overall recovery capability of the company.

The exercise can be counted as a success if the participants leave feeling that they have learned something and that next time they know how they can do better.

No one ever fails, although there may be a few red faces from time to time.

EXERCISE PROGRAMMES

Testing is accepted as being a cost of both time and funds and therefore it is wise to consider a test programme that demonstrates a sufficient level of confidence to serve both the proving and rehearsal aspects discussed above.

Technical proving should be done at least once a year in accordance with the contract terms of the standby service. This regular proving is even more important if the standby service is an internal resource that may be diverted for good business reasons and may not actually be there on the day

2. Evaluation Criteria for Business Continuity Plans

that you need it. Third party contracts are usually more secure in that the supplier's business will depend on the success of the service that they provide.

Rehearsal sessions for the senior managers are usually more difficult to establish owing to the limited availability in their full diaries. It is well worth considering short exercises, run as tabletop discussions, in conjunction with other regular staff meetings, such as board meetings. There is often as much value in a one-hour exploration of a specific, hypothetical incident as there would be in using a full day to trawl through a more complex scenario. Even if only a few senior managers are available, the resulting embarrassment or otherwise will be discussed and passed around the board room and the lessons learned will be to the general good of the company plan.

It is also worth considering when such rehearsal sessions should be run and what to expect from them at each stage of the business continuity management development.

Exercises run very early in the development programme will usually reinforce the need for senior management support throughout the programme. Exercises run during the later stages will help to refine the business continuity plan documentation and weed out expressions and procedures that are unacceptable. Exercises at the end of the implementation will enable managers to practice their emergency roles.

Step 9
Public Relations and Crisis Co-ordination

Develop, co-ordinate, evaluate, and exercise plans to handle the media during crisis situations. Develop, co-ordinate, evaluate, and exercise plans to communicate with and, as appropriate, provide trauma counselling for employees and their families, key customers, critical suppliers, owners/ stockholders, and corporate management during crisis. Ensure all stakeholders are kept informed on an as-needed basis.

Corporate governance and good stewardship lend an increased emphasis to the need to provide a practised response to the media, especially in cases where the public reputation of the organisation may be prejudiced by the lack of a well considered press release.

There is also an increasing expectation that the organisation must take responsibility for the well being of all staff, a function that is often initially discharged through various forms of personal counselling.

MEDIA MANAGEMENT

The business world is increasingly vulnerable to adverse criticism in the media where there is always an avid desire to emphasise the down side of any story, especially when there is the possibility of allocating blame to someone. The notion of corporate culpability is here to stay and the threat of litigation is ever present. When disaster strikes, corporate executives must be prepared to manage the media interest from the outset. There are three things to consider:

- limit the interface with the media to nominated and trained people;
- prohibit casual interviews;
- make use of pre-prepared statements to the press where the wording can be thought through in advance and all sensitivities can be addressed.

CRISIS COMMUNICATION

Crisis communication should be considered for both internal and external recipients. If the stricken company is part of a group, then it will be as important to control the messages that pass around to other less well-informed sister units as it will to control the messages that reach customers and competitors.

Again there is a short list of principle considerations:

- use pre-prepared texts to be sent out in advance of public speculation. Admit the truth and emphasise the planned recovery programme;
- ensure that important customers receive regular updates of your progress back to normality. Customers will not go elsewhere if they know when to expect you to be working again;
- emphasis the positive aspects of your information security preparations and the benefits of your planned recovery procedures.

TRAUMA COUNSELLING

At the moment when disaster strikes, be it fire, flood or famine, the impact on the people involved can be masked by their immediate reactions to the event. Often the deeper indications of concern take much longer to emerge. For example, the rate of staff turnover increased dramatically in the six months following the terrorist bombs in London for those companies who were materially affected by the damage. It is thought that this was due to little more than 'bad feelings' about the work place and an increased desire to change, even if the change was only to the building next door. Several companies who survived the bomb damage were hit again by high staff turnover when they could have done more to recognise the problem.

Trauma counselling is one way to establish a close communication with individual members of staff in a way that demonstrates care for the individual and improves the feeling of vulnerability to the disaster.

Step 10
Co-ordination with Public Authorities

Establish applicable procedures and policies for co-ordinating response, continuity, and restoration activities with local authorities while ensuring compliance with applicable statutes or regulations.

In larger incidents the involvement of police, fire and ambulance services must be anticipated. Organisations will be expected to co-operate with these services where they are involved in the emergency. This co-operation may also extend to include the Local Authority emergency planning officer (EPO) if the scope of the incident affects neighbouring premises or general public areas.

The public services are just that, publicly funded and publicly tasked to protect the environment and individual members of the public. Companies should not expect the police to become a private security force in the aftermath of a major incident. Once the event is under control and made safe for the public it will be up to the company to resume control of the physical security of the site. Often reports of looting at the scene of major disasters come about merely through the lack of continuity between the public services and the owners of the building at that point of hand-over.

It is important to understand the roles of each of the 'Blue Light' services, for example, at the scene of a major fire. In general terms:

2. Evaluation Criteria for Business Continuity Plans

- the fire service will control what they determine to be the 'fire ground'. This is the area where there is a significant danger to life and limb from the fire and or its immediate or potential fuel supply;
- the police will establish a cordon outside the fire ground as a further control over the public and to ensure that the service vehicles can obtain adequate access to the site;
- the ambulance service will remain outside the police cordon, except for any paramedic requirement, which will come under the temporary command of the senior fire officer present.

Once the fire service deems the site to be safe, they will leave the site and the police will determine if there is still a danger to the public. If not, they will hand over responsibility to the nominated key holder if present. If the key holder is not present they will monitor the security of the building in accordance with their normal rounds, but they will not accept responsibility for its security or its contents.

If there is a suspicion that the fire has been caused by arson, the scene becomes a crime scene and access to the site will be limited to 'Scene of Crime Officers' (SOCO). If there has been a significant fire, it may be several days before the building cools down sufficiently for a full forensic search to be carried out and access to the site will be severely limited until that activity has been completed.

If your fire safe, now buried in the building, contains your only software copies, it may be days before you are allowed to recover them. It pays to establish a good liaison with the SOCO and their team at the earliest opportunity.

If the incident results in a lasting danger to the public, such as a chemical spillage that may take days to clear up or

2. Evaluation Criteria for Business Continuity Plans

threatens other businesses or residents in the area, you can expect the Local Authority to take control of the longer-term management of the problem. Each Local Authority has an EPO who would work with the Health and Safety Executive to oversee the necessary clean up measures.

The EPO will provide a useful source of expertise and resources from the skills and services maintained by the local council. It pays to establish a good liaison with the EPO and their team at the earliest opportunity.

If the damage is significant, then you are heading for an insurance claim of some sort. And, given that the insurance claim is going to be for a significant sum, you can expect the insurance company to appoint a loss assessor to protect their interests. The appointment of a salvage contractor as part of the business continuity plan is one way to ensure that a competent professional advisor is on hand within hours of the incident being declared, to protect your interests. Salvage contractors are also familiar with the rules and procedures as required by the emergency services and because of their own training and expertise, will often be the first to be authorised to enter a damaged site, to retrieve your back-up tapes!

Appendix A
General Business Continuity Terms

Like all subjects business continuity planning uses a variety of terms and expressions to explain the scope of its material. Business continuity planning is also a comparatively new subject and so it is worth taking time to read through the following glossary of terms for ease of reference.

Activation
The implementation of recovery procedures, activities and plans in response to an emergency or disaster declaration.

Alternative site
An alternative operating location for the usual business functions (i.e. support departments, information systems and manufacturing operations) when the primary facilities are inaccessible. Associated term: back up site.

Alert
A formal notification that an incident has occurred which may develop into a disaster.

BS7799
A UK BSI Standard for information security management. Section 9 deals with business continuity management.

Backlog trap
The effect on the business of a backlog of work that develops when a system or process is unavailable for a long period, and which may take a considerable length of time to reduce.

Appendix A. General Business Continuity Terms

Building denial
Any damage, failure or other condition which causes denial of access to the building or the working area within the building, e.g. fire, flood, contamination, loss of services, air conditioning failure, forensics.

Business continuity
A proactive process which identifies the key functions of an organisation and the likely threats to those functions, from this information plans and procedures which ensure key functions can continue whatever the circumstances can be developed.

Business continuity co-ordinator
A member of the recovery management team who is assigned the overall responsibility for co-ordinator of the recovery planning programme ensuing team member training, testing and maintenance of recovery plans. Associated terms: business recovery planner, disaster recovery planner, business recovery co-ordinator, disaster recovery administrator.

Business continuity plan (BCP)
A collection of procedures and information which is developed, compiled and maintained in readiness for use in the event of an emergency or disaster. Associated terms: business recovery plan, disaster recovery plan, recovery plan.

Business continuity management (BCM)
Those management disciplines, processes and techniques which seek to provide the means for continuous operation of the essential business functions under all circumstances.

Business continuity planning
The advance planning and preparations which are necessary to identify the impact of potential losses; to formulate and implement viable recovery strategies; to develop recovery plan(s) which ensure continuity of organisational services in

the event of an emergency or disaster; and to administer a comprehensive training, testing and maintenance programme. Associated terms: contingency planning, disaster recovery planning, business recovery planning.

Business continuity programme
An ongoing process supported by senior management and funded to ensure that the necessary steps are taken to identify the impact of potential losses, maintain viable recovery strategies and recovery plans, and ensure continuity services through personnel training, plan testing and maintenance. Associated terms: disaster recovery programme, business recovery programme, contingency planning programme.

Business critical point
The latest moment at which the business can afford to be without a critical function or process.

Business impact analysis (BIA)
A management level analysis which identifies the impacts of losing company resources. The BIA measures the effect of resource loss and escalating losses over time in order to provide senior management with reliable data upon which to base decisions on risk mitigation and continuity planning. Associated terms: business impact assessment, business impact analysis assessment.

Cold site
One or more data centres or office space facilities equipped with sufficient pre-qualified environmental conditioning, electrical connectivity, communications access, configurable space and access to accommodate the installation and operation of equipment by critical staff required to resume business operations.

Appendix A. General Business Continuity Terms

Contingency fund
An operating expense that exists as a result of an interruption or disaster which seriously affects the financial position of the organisation. Associated term: extraordinary expense.

Contingency plan (a general non-specific point)
A plan of action to be followed in the event of a disaster or emergency occurring which threatens to disrupt or destroy the continuity of normal business activities and which seeks to restore operational capabilities.

Crisis
An abnormal situation, or perception, which threatens the operations, staff, customers or reputation of an enterprise.

Crisis management team (CMT)
A group of executives who direct the recovery operations whilst taking responsibility for the survival and the image of the enterprise.

Crisis plan or Crisis management plan
A plan of action designed to support the crisis management team when dealing with a specific emergency situation which might threaten the operations, staff, customers or reputation of an enterprise.

Critical service
Any service which is essential to support the survival of the enterprise.

Critical data point
The point to which data must be restored in order to achieve recovery objectives.

Decision point
The latest moment at which the decision to invoke emergency procedures has to be taken in order to ensure the continued viability of the enterprise.

Declaration (of disaster)
A formal statement that a state of disaster exists.

Disaster
Any accidental, natural, or malicious event which threatens or disrupts normal operations, or services, for sufficient time to affect significantly, or to cause failure of, the enterprise.

Disaster recovery plan (DRP) or Recovery plan
A plan to resume, or recover, a specific essential operation, function or process of an enterprise.

Disaster recovery (DR)
The process of returning a business function to a state of normal operations either at an interim minimal survival level and/or re-establishing full scale operations.

Emergency data services
Remote capture and storage of electronic data, such as journalling, electronic vaulting and database shadowing.

Emergency
An actual or impending situation that may cause injury, loss of life, destruction of property or interfere with normal business operations to such an extent to pose a threat of disaster.

Emergency control centre
The location from which disaster recovery is directed and tracked; it may also serve as a reporting point for deliveries, services, press and all external contacts.

Emergency management team
The group of staff who command the resources needed to recover the enterprise's operations.

APPENDIX A. GENERAL BUSINESS CONTINUITY TERMS

Emergency management plan
A plan which supports the emergency management team by providing them with information and guidelines.

Enterprise
An organisation, a corporate entity; a firm, an establishment, a public or government body, department or agency; a business or a charity.

Enterprise (large scale or super)
An enterprise that is large and complex, in the sense that it could absorb the impact of losing a complete location or business unit. The normal terminology, and perspective, needs to be scaled down by regarding individual locations or business units as self-sustaining entities.

Financial impact
An operating expense that continues following an interruption or disaster, which as a result of the event cannot be offset by income and directly affects the financial position of the organisation.

Hot site
A data centre facility or office facility with sufficient hardware, communications interfaces and environmentally controlled space capable of providing relatively immediate backup data processing support. Associated terms: warm site, cold site.

Human Resource Disaster Recovery
(HRDR) A specific strategy for dealing with risk assessment, prevention, control and business recovery for critical (key) personnel.

Immediate recovery team
The team with responsibility for implementing the business continuity plan and formulating the organisation's initial recovery strategy.

Impact
Impact is the cost to the enterprise, which may or may not be measured in purely financial terms.

Incident
Any event which may be, or may lead to, a disaster.

Invocation
A formal notification to a service provider that its services will be required.

Information security
The securing or safeguarding of all sensitive information, electronic or otherwise, which is owned by an organisation.

Logistics/Transportation team
A team comprised of various members of departments associated with supply acquisition and material transportation, responsible for ensuring the most effective acquisition and mobilisation of hardware, supplies and support materials.

Mobile standby
A transportable operating environment, usually complete with accommodation and equipment, which can be transported and set up at a suitable site at short notice.

Mobilisation
The activation of the recovery organisation in response to an emergency or disaster declaration.

Off-site location
A storage facility at a safe distance from the primary facility which is used for housing recovery supplies, equipment, vital records, etc.

Operational impact
An impact which is not quantifiable in financial terms but its effects may be among the most severe in determining the survival of an organisation following a disaster.

Outage
The interruption of automated processing systems, support services or essential business operations which may result in the organisation's inability to provide service for some period of time.

Period of tolerance
The period of time in which an incident can escalate to a potential disaster.

Pre-positional resource
Material (i.e. equipment, forms and supplies) stored at an off-site location to be used in business resumption and recovery operations. Associated terms: pre-positioned inventory.

Reciprocal agreement
An agreement in which two parties agree to allow the other to use their site, resources or facilities during a disaster.

Recovery
See system recovery.

Recovery exercise
An announced or unannounced execution of business continuity plans intended to implement existing plans and/or highlight the need for additional plan development. Associated terms: disaster recovery test, disaster recovery exercise, recovery test, recovery exercise.

Recovery management team
A team of people, assembled in an emergency, who are charged with recovering an aspect of the enterprise, or obtaining the resources required for the recovery.

Recovery plan
A plan to resume a specific essential operation, function or process of an enterprise. Traditionally referred to as a disaster recovery plan (DRP).

Recovery site
A designated site for the recovery of computer or other operations, which are critical to the enterprise.

Recovery strategy
A pre-defined, pre-tested, management approved course of action to be employed in response to a business disruption, interruption or disaster.

Recovery team
A group of individuals given responsibility for the co-ordination and response to an emergency or recovering a process or function in the event of a disaster.

Recovery window
The time scale within which time sensitive function or business units must be restored, usually determined by means of a business impact analysis.

Resilience
The ability of a system or process to absorb the impact of component failure and continue to provide an acceptable level of service.

Response
The reaction to an incident or emergency in order to assess the level of containment and control activity required.

Restart
The procedure or procedures that return applications and data to a known start point. Application restart is dependent upon having an operable system.

Restoration
The process of planning for and implementing full scale business operations which allow the organisation to return to a normal service level.

Appendix A. General Business Continuity Terms

Resumption
The process of planning for and/or implementing the recovery of critical business operations immediately following an interruption or disaster.

Risk assessment and management
The identification and evaluation of operational risks that particularly affect the enterprise's ability to function and addressing the consequences.

Risk reduction or mitigation
The implementation of the preventative measures which risk assessment has identified.

Scenario
A pre-defined set of events and conditions which describe an interruption, disruption or disaster related to some aspect(s) of an organisation's business for purposes of exercising a recovery plan(s).

Security review
A periodic review of the security of tangible and intangible assets which should cover security policy, effectiveness of policy implementation, restriction of access to the assets, accountability for access and basic safety.

Service level agreement (SLA)
An agreement between a service provider and service user as to the nature, quality, availability and scope of the service to be provided.

Site access denial
Any disturbance or activity within the area surrounding the site which renders the site unavailable, e.g. fire, flood, riot, strike, loss of services, forensics. The site itself may be undamaged.

Appendix A. General Business Continuity Terms

Social impact
Any incident or happening that affects the well-being of a population and which is often not financially quantifiable.

Standby service
The provision of the relevant recovery facilities, such as cold site, warm site, hot site and mobile standby.

Stand down
Formal notification that the alert may be called off or that the state of disaster is over.

Structured walk-through
An exercise in which team members verbally review each step of a plan to assess its effectiveness and identify enhancements, constraints and deficiencies. Associated term: bench test.

System denial
A failure of the computer system for a protracted period, which may impact an enterprise's ability to sustain its normal business activities.

System recovery
The procedures for rebuilding a computer system to the condition where it is ready to accept data and applications. System recovery depends on having access to suitable hardware.

System restore
The procedures that are necessary to get a system into an operable condition where it is possible to run the application software against the available data. System restore depends upon having a live system available.

Table top exercise
The exercising and testing of a business continuity plan, using a range of scenarios whist not effecting the enterprise's normal operation.

Appendix A. General Business Continuity Terms

Tolerance threshold
The maximum period of time which the business can afford to be without a critical function or process.

Vendor
An individual or company providing a service to a department or the organisation as a whole. Associated terms: supplier, third party vendor.

Vital record
A record that it is essential for preserving, continuing or reconstructing the operations of the organisation and protecting the rights of the organisation, its employees, its customers and its stockholders.

Warm site
A data centre or office facility which is partially equipped with hardware, communications interfaces, electricity and environmental conditioning capable of providing backup operating support. Associated terms: hot site, cold site.

Work area standby
A permanent or transportable office environment, complete with appropriate office infrastructure.

This glossary has been compiled by the BCI and may only be reproduced under the following conditions:

a) The BCI is informed of the intention to reproduce;
b) The Glossary is reproduced in its entirety;
c) The source of the material is accredited to the BCI.

Appendix B
Additional Sources of Information

The following organisations are focal points for additional information on business continuity planning. Each organisation maintains an active web site that is regularly updated and will be a reliable source of current information long after the ink is dry on this page.

The Business Continuity Institute
www.thebci.org

The Disaster Recovery Institute International
www.dr.com

Survive Business Continuity Group
www.survive.com

Acknowledgements

The publishers would like to thank the Business Continuity Institute for their contribution to this book and for their permission to reproduce the Glossary of Terms. And to Dr Tony Burns-Howell and Andrew Seymour for their advice on an earlier draft of this text.

Commissioning editor: Dr Martin Gill, Scarman Centre, Leicester University, 154 Upper New Walk, Leicester, LE2 7QA, UK. (mg26@le.ac.uk).